**Association of Ukrainian grant holders DAAD**
**AUS DAAD**
**National Committee IAESTE Ukraine**
**NC IAESTE-Ukraine**

**Series: "Modern Mathematics for Engineers"**

*Tamara G. Stryzhak*

# Minimax criteria of stability

$$L = T - \Pi$$

$$\max_{q}\left\langle \min_{\dot{q}} L\left(t, \dot{q}_j, q_j\right)\right\rangle$$

Tamara G. Stryzhak

# THE MINIMAX CRITERION FOR STABILITY

*ibidem*-Verlag
Stuttgart

**Bibliografische Information der Deutschen Nationalbibliothek**
Die Deutsche Nationalbibliothek verzeichnet diese Publikation in der Deutschen Nationalbibliografie; detaillierte bibliografische Daten sind im Internet über http://dnb.d-nb.de abrufbar.

**Bibliographic information published by the Deutsche Nationalbibliothek**
Die Deutsche Nationalbibliothek lists this publication in the Deutsche Nationalbibliografie; detailed bibliographic data are available in the Internet at http://dnb.d-nb.de.

∞

Gedruckt auf alterungsbeständigem, säurefreien Papier
Printed on acid-free paper

ISBN-10: 3-89821-919-4

ISBN-13: 978-3-89821-919-8



Printed in Germany

I·A·E·S·T·E

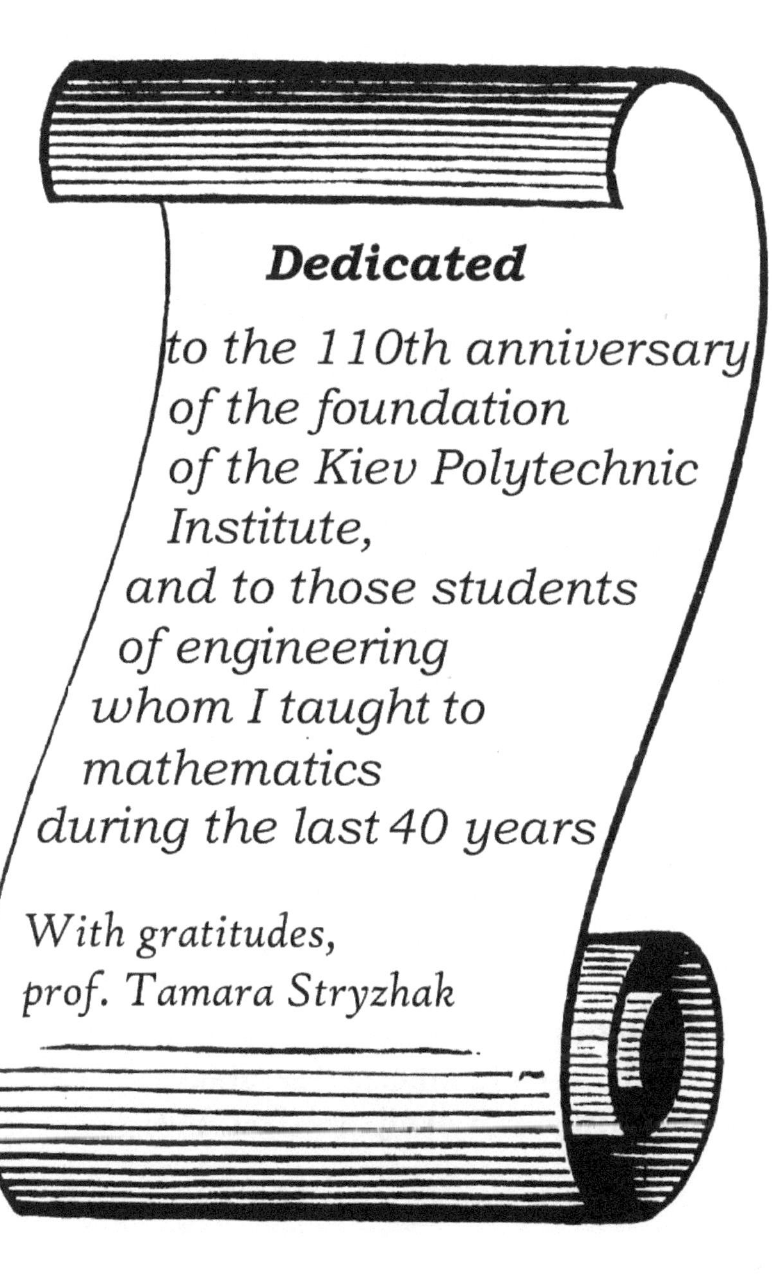
**Dedicated**
*to the 110th anniversary*
*of the foundation*
*of the Kiev Polytechnic*
*Institute,*
*and to those students*
*of engineering*
*whom I taught to*
*mathematics*
*during the last 40 years*
*With gratitudes,*
*prof. Tamara Stryzhak*

**Series: "Modern Mathematics for Engineers"**

Lectures for the trainee-students of IAESTE

**Educational publication**

*Lectures*
**Professor NTUU "KPI" Tamara Stryzhak**

*Translator*
**Faculty member NTUU "KPI" Nataliya Sarycheva**
*English language editors*
**IAESTE trainee students:**
**David Ellis (United Kingdom)**
**Sarah Burden (United Kingdom)**
**Andrea Trautsamwieser (Austria)**
**Erkan Koc (Germany)**
**Leonard Neuhaus (Germany)**
**Eunice Kim (USA)**
**Heide Gieber (Austria)**
**Peter Smith (United Kingdom)**
**Betina Carina Sieber (Germany)**
**Laura Lambertz (Germany)**
**Mirjana Vukelja (Swizerland)**

**Series: "Modern Mathematics for Engineers"**

# Minimax criteria of stability

**Professor NTUU "KPI"**
**Tamara Stryzhak**

*Lectures*

This publication presents the “Minimax criterion of stability”. The “Minimax criterion of stability” is a criterion sufficient for determining the motion stability of a mechanical system under the influence of small-amplitude high-frequency parameter oscillations. This criterion allows the calculation of stability conditions of the oscillations of a non-stable mechanical system. It does not involve the use of motion equations, rather, it uses only the Langrangian function L= T – Π.

All remarks and recommendations are greatly appreciated and will be taken into considerations in subsequent editions.

Phone: +380 44 2418648
Fax: +380 44 2418648, +380 44 2417620
E-mail: stri@aer.ntu-kpi.kiev.ua
Web-site: www.iaeste.org.ua

# Contents

# Introduction

## (a) Acknowledgements

> "Mathematical Truth stays put in centuries,
> Metaphysical Ghosts disappear
> like delirium of the sick."
> *Voltaire*

*Minimax Criterion of Stability* is the first volume within the project *Modern Mathematics for Engineers*.

The project *Modern Mathematics for Engineers* was launched by AUS–DAAD and the National Committee IAESTE Ukraine. The purpose of the project is to publish the most essential mathematical results, to present them with enough clarity for engineers to apply and, ante omnia, to realize the scientific exchange in the sphere of applied mathematics, the oscillation theory, theoretical mechanics, etc.

We expect to realize this purpose first of all with the help of IAESTE trainee students.

The first trainee students who took part in the project:

- David Ellis (Imperial College of London, Great Britain);
- Andrea Trautsamwieser (Technical University of Vienna, Austria);
- Sarah Burden (University of St. Andrews, UK);
- Leonard Neuhaus (Ludwig Maximilians University of Munich, Germany);
- Erkan Koc (University of Bonn, Germany);
- Henning Hans Petzka (University of Aahen, Germany);
- Eunice Kim (Carnegie Mellon University, USA);
- Peter Smith (Queen's University Belfast, UK);
- Heide Gieber (Vienna University of Technology, Austria);
- Laura Lambertz (RWTH Aachen, Germany);

- Mirjana Vukelja (ETH Zurich, Switzerland);
- Bettina Carina Sieber (TU Munich, Germany).

We invite for cooperation all mathematicians and engineers who are interested in having their scientific works published in the frame of the project.

We were inspired to launch the project by the positive experience of the researchers of the Californian University, who published the monograph *Modern Mathematics for Engineers*[1] more than half a century ago, in 1956. This work was very successful. As a matter of fact this monograph laid the solid foundation for the applied mathematics to develop successfully and steadily. This monograph is an excellent example to follow. Thus, like 50 years ago, we are launching the project with researches dedicated to the pendulum. We can hardly remember any other mechanism which is simpler than the pendulum, the mechanism whose scientific life has been so rich in application in different spheres of organic and inorganic nature, as the pendulum has lived a long and rich in discoveries life. I think it is the right time to build a monument to a simple and meaningful device.

Our contribution into realization of this project included taking the following steps:

1) publishing the monograph *Research Methods of the Pendulum Dynamic Systems* [4];
2) receiving "Minimax Criterion of Stability" and applying it in order to research a number of mathematical models and proved, in particular, that any position of the pendulum, even a horizontal one in the vertical plane, can be made stable with the help of the suspension point oscillations;
3) building an installation to demonstrate our theoretical results.

All in all we tried our best to state the results avoiding proving theorems as their translation from Russian into English might have some inadequacy as grammars of these languages have a certain level of ambiguity. Instead of proving the theorems we shall follow Lopital's words, the author of the first textbook on Mathematical Analysis, and say, "We pass the word of honor that the theorem is true".

1 Edited by Edwin F. Beckenbach (Professor of Mathematics, University of California, Los Angeles). New York, Toronto, London: McGraw-Hill 1956.

## (b) Oscillations – what are they?

"Nature prefers oscillating motions in all demonstrations of life. Not without reason we can assume that there is some optimality property at the back of this phenomenon."[8]

It would be rather difficult to mention all publications about the Pendulum. As a matter of fact, oscillations of pendulum systems were researched by Galileo, Newton, Euler, Huygens and many other scientists who made a great contribution into studying the mechanism of the pendulum motion, which was of great importance for the history of mankind's discoveries and technical progress. The pendulum has been used in clocks to define time, in special devices to measure the terrestrial gravitation, as a plumb line in building to define the vertical, etc. The Earth's rotation was proved with the help of the pendulum oscillations as well.

Pendulum systems are, as a rule, non-linear and require specific methods of research. For instance, creation of the elliptical function theory by Abel, Jacoby, and Weierstrass is also connected with the research of the mathematical pendulum oscillations.

Nowadays due to the mathematicization of research in different sciences there has been an increasing interest in studying motions of pendulum systems. The following special terms have appeared: the pendulum law of the population migration, the pendulum law of the rhythm regulator action, the pendulum of emotions, etc.

It is a well-known fact that all living organisms have so called biological clocks, at the basis of which there is an oscillating system – a non-linear oscillator. It is well known that the vestibular apparatus of animals and humans contains three non-linear pendulum systems, which are located in three mutually perpendicular planes. Oscillations are present everywhere: in the opening of a flower at the sunrise, in the growth of an embryo, in germinating of a grain, in the heart beating, in the work of a dental drill and a jackhammer, in the rise and fall of the tide. Besides, wherever there is life, there are oscillations at the cellular level.

The nature of forces which cause oscillations is variable, but the result of these forces is the same, namely - oscillations.

This information bears a descriptive character as we would like to attract the readers' attention to the oscillation theory to convince them of the fact that oscillation processes imply something much deeper than what we actually know: that the source of the oscillating processes is a hidden potential force and kinetic energyOscillations of a non-linear oscillator can be forced or can have a free-running character (such as the heart or aorta beating, biorhythms, etc.). In fact, the pendulum motion laws, periodic or almost-periodic oscillations, are immanent in the whole physical world.

A human being recgives the main information about the outer world via sound and light oscillations, analyzing which scientists use pendulum systems. The latter are found in different engineering tasks. For example, oscillations in electrical and mechanical systems, rocking of ships on water, oscillations of satellites, vibration of the hull and the wings of planes, movements of cables, chains, travelling or gantry cranes, etc. All these phenomena are explained with analogous differential equations which describe motions.

At this point our introduction is completed and we turn to the beautiful algorithmic mathematical language: *"A" is given, "B" is to be proved.*

## (c) Minimax criteria of stability

The stability of oscillations of a mechanical or electrical system, which is affected by low amplitude high-frequency disturbances, is researched. It is shown that the affect of vibration can be replaced by the action of some potential force. Under the action of some disturbance the stable equilibrium position can become unstable, and the unstable equilibrium position can become stable. New positions of stable dynamic equilibrium can also appear. Conclusions about the stability can be made with the help of the minimax criteria for stability described in this work. The criteria for stability rests on the time averaging operation of the canonical system of differential equations. The minimax criteria for stability is applied when the frequency of external disturbance is substantially higher than the natural vibration of the quiescent system.

In this work the minimax criteria for stability is applied in the research of the stability of the pendulum systems equilibrium position. In particular, it is found that any position of the pendulum, even a horizontal one, can be made stable via vibrations of the pendulum's suspension point.

An experimental installation to check the received theoretical conclusions was created. All the conclusions were proved correct.

This work illustrates simple examples of the minimax criteria for stability application. It also describes the experiment. At the end of this work the mathematical survey of the minimax criteria for stability is presented.

## §1 Minimax stability criteria definition

Here the movement of a mechanical or electrical system is studied. We shall define the generalized coordinates by $q_j$ $(j=1,...,n)$, the generalized velocities by $\dot{q}_j$ $(j=1,...,n)$ and the time by $t$.

Let $L=L(t,\dot{q}_j,q_j)$ be the kinetic potential, where $L=\mathrm{T}-\Pi$, and $\mathrm{T}$ is the kinetic energy of the system and $\Pi$ is the potential energy. We shall find the minimum of the kinetic potential $L$ as a function of the generalized velocities $\dot{q}_j$,

$$\Pi_0(t,q_j)=\min_{\dot{q}_s} L(t,\dot{q}_j,q_j), \quad (j,s=1,...,n). \tag{1}$$

We shall exclude from $L(t,\dot{q}_j,q_j)$ the derivative $\dot{q}_j$ by the necessary conditions of extremum

$$\frac{\partial L(t,\dot{q}_j,q_j)}{\partial \dot{q}_s}=0, \quad (j,s=1,...,n). \tag{2}$$

Using the operation of time averaging $t$ on the function $\Pi_0(t,q_j)$ we are able to eliminate $t$

$$\Pi_0(q_j)\equiv\langle\Pi_0(t,q_j)\rangle\equiv\lim_{\mathrm{T}\to+\infty}\frac{1}{\mathrm{T}}\int_0^{\mathrm{T}}\Pi_0(t,q_j)\,dt. \tag{3}$$

If the function $\Pi_0(q_j)$ has a maximum at the point $q_j=q_{j0}$ $(j=1,...,n)$, then the maximum corresponds to the stable position of the dynamic equilibrium of the initial oscillating system. Meanwhile the system is steadily oscillating at the position $q_j=q_{j0}$ $(j=1,...,n)$.

To find the position of the stable state it is first necessary to find the minimum of the function $L(t,\dot{q}_j,q_j)$ via variable $\dot{q}_j$, and then find the maximum of the function using,

$$\max_{q_j}\left\langle \min_{\dot{q}_j} L(t,\dot{q}_j,q_j)\right\rangle,$$

by which the name *'minimax criteria of stability' arises.*

It is important to mention that the conditions of stability are found without using Lagrange's differential equations [1]

$$\frac{d}{dt}\frac{\partial L}{\partial \dot{q}_s}-\frac{\partial L}{\partial q_s}=0, \quad (s=1,\ldots,n)$$

and only the knowledge of Lagrange's function is used $L=L(t,\dot{q}_j,q_j)$.

## §2 Stability of the mathematical pendulum oscillations at the upper equilibrium position

We shall consider oscillations at the upper equilibrium position of the mathematical pendulum with length $l$ and mass $m$, with the vibrating vertical point of support $A$. We shall define the coordinates of the centre of gravity of the pendulum by $x, y$, the pendulum angle from the vertical by $\varphi$ and the deflection of the point $A$ of the pendulum support in a vertical direction (figure 1) by $r(t) = a\sin\omega t$, where $a$ is the amplitude and $\omega$ is the frequency of the vibrations.

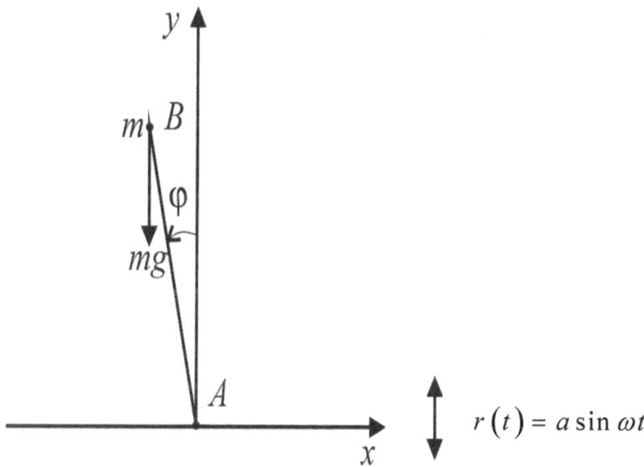


Figure 1

For the coordinates of the centre of gravity we find the following,

$$x = -l\sin\varphi, \quad y = l\cos\varphi + r,$$

and for the velocities

$$\dot{x} = -l\cos\varphi\cdot\dot{\varphi}, \quad \dot{y} = -l\sin\varphi\cdot\dot{\varphi} + \dot{r}.$$

The kinetic and potential energies are found to have the following forms

$$\mathrm{T} = \frac{m}{2}\left(\dot{x}^2 + \dot{y}^2\right) = \frac{m}{2}\left(l^2\dot{\varphi}^2 - 2l\dot{\varphi}\sin\varphi\cdot\dot{r} + \dot{r}^2\right),$$

$$\Pi = mg\left(l\cos\varphi + r\right).$$

Then the Lagrange function is as follows,

$$L = \mathrm{T} - \Pi = \frac{m}{2}\left(\left(l\dot{\varphi} - \dot{r}\sin\varphi\right)^2 + \dot{r}^2\cos^2\varphi\right) - mg\left(l\cos\varphi + r\right) \qquad (4)$$

and $\min\limits_{\dot{\varphi}} L$ is found with the help of the equation

$$\frac{\partial L}{\partial\dot{\varphi}} = m\left(l\dot{\varphi} - \dot{r}\sin\varphi\right)\cdot l = 0$$

to give the following expression

$$\min_{\dot{\varphi}} L = \frac{m}{2}\dot{r}^2\cos^2\varphi - mgl\cos\varphi - mgr \qquad (5)$$

As $r(t) = a\sin\omega t$ we get

$$\langle r\rangle = \langle a\sin\omega t\rangle = 0, \qquad \left\langle\dot{r}^2\right\rangle = \left\langle a^2\omega^2\cos^2\omega t\right\rangle = \frac{a^2\omega^2}{2}$$

where <X> is the time average of the function X.

Therefore the following is found,

$$\Pi_0(\varphi) = \left\langle\min_{\dot{\varphi}} L\right\rangle = mgl\cos\varphi - \frac{m}{2}\frac{a^2\omega^2}{2}\cos^2\varphi,$$

where $-\Pi_0(\varphi)$ is a dynamic analogue of potential energy.

The function $-\Pi_0(\varphi)$ has its minimum at the point $\varphi = 0$, if the following condition is satisfied

$$\frac{\partial^2 \Pi_0(\varphi)}{\partial \varphi^2} < 0 \quad \text{or} \quad mgl - \frac{m}{2} a^2 \omega^2 < 0 .$$

The stability criteria is reduced to the following inequality,

$$a^2 \omega^2 > 2gl . \qquad (6)$$

Relation (6) was discovered earlier in relevant scientific work. It is obvious that it was first discovered in some works [2, 3], and later repeated in the works of N. Bogolyubov, P. Kapitsa, G. Stoker, K.Valeev, T. Stryzhak.

## §3 Optional oscillations of the pendulum's point of support

Let the pendulum's supporting point move in vertical and horizontal directions in the following manner

$$x_A = s(t), \quad y_A = r(t),$$

where $s(t)$, $r(t)$ are periodic or quasi-periodic functions. Point B has the following coordinates,

$$x = s(t) - l\sin\varphi, \quad y = r(t) + l\cos\varphi.$$

The pendulum's kinetic and potential energy are defined by the following expressions

$$\mathrm{T} = \frac{m}{2}\left(l^2\dot{\varphi}^2 - 2l\dot{\varphi}\left(\dot{s}\cos\varphi + \dot{r}\sin\varphi\right) + \dot{s}^2 + \dot{r}^2\right)$$

$$\Pi = mg\left(l\cos\varphi + r\right).$$

The Lagrangian function therefore has the following form,

$$L = \frac{m}{2}\left(\left(l\dot{\varphi} - \dot{s}\cos\varphi - \dot{r}\sin\varphi\right)^2 + \left(\dot{s}\sin\varphi - \dot{r}\cos\varphi\right)^2\right) - mgr - mgl\cos\varphi.$$

Now suppose that $\langle r\rangle = 0$ to find

$$\min_{\dot{\varphi}} L = \frac{m}{2}\left(\dot{s}\sin\varphi - \dot{r}\cos\varphi\right)^2 - mgr - mgl\cos\varphi$$

and

$$\Pi_0(\varphi) \equiv \left\langle \min_{\dot\varphi} L \right\rangle = \frac{m}{2}\left(\langle \dot s^2 \rangle \sin^2\varphi - 2\langle \dot s \dot r \rangle \sin\varphi \cos\varphi + \langle \dot r^2 \rangle \cos^2\varphi - 2gl\cos\varphi\right).$$

The pendulum has the vertical equilibrium position $\varphi = 0$, if $\Pi_0'(0) = 0$, i.e. at $\langle \dot s \dot r \rangle = 0$. Position $\varphi = 0$ will be stable if $\Pi_0''(0) < 0$, i.e.

$$2\langle \dot s^2 \rangle - 2\langle \dot r^2 \rangle + gl < 0$$

or

$$\langle \dot r^2 \rangle > \frac{gl}{2} + \langle \dot s^2 \rangle.$$

Thus, the vertical oscillation of the supporting point improves the stability of the upper vertical position of the pendulum, and the horizontal oscillations of the supporting point reduce the stability of the upper vertical position of the pendulum.

Now we shall consider the stability of a random position $\varphi = \varphi_0$ of the mathematical pendulum. At that position, $\varphi = \varphi_0$ will be the equilibrium position, if $\Pi_0'(\varphi_0) = 0$, i.e.

$$\left(\langle \dot s^2 \rangle - \langle \dot r^2 \rangle\right)\sin 2\varphi_0 - 2\langle \dot s \dot r \rangle \cos 2\varphi_0 + 2l\,\mathrm{g}\sin\varphi_0 = 0 \qquad (7)$$

The position $\varphi = \varphi_0$ will be stable if

$$\left(\langle \dot s^2 \rangle - \langle \dot r^2 \rangle\right)\cos 2\varphi_0 + 2\langle \dot s \dot r \rangle \sin 2\varphi_0 + l\,\mathrm{g}\cos\varphi_0 < 0 \qquad (8)$$

Let, for instance, the pendulum's supporting point oscillate in a harmonic manner with an amplitude $a$ along a straight line which makes an angle $\theta$ with the vertical (figure 2).

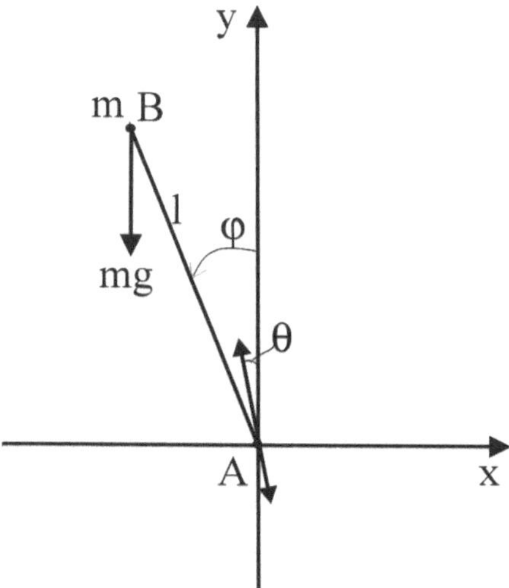


Figure 2

Here the coordinates of point $B$ are as follows,

$$x_B = -a\sin\theta \cdot \sin\omega t - l\sin\varphi\,,\ \ y_B = a\cos\theta \cdot \sin\omega t + l\cos\varphi\,.$$

and the following expressions are found

$$s(t) = -a\sin\theta\sin\omega t\,,\ \ r(t) = a\cos\theta\sin\omega t$$

$$\left\langle \dot{s}^2 \right\rangle = \frac{a^2\omega^2\sin^2\theta}{2},\ \ \left\langle \dot{s}\dot{r} \right\rangle = -\frac{a^2\omega^2}{2}\sin\theta\cos\theta\,,\ \ \left\langle \dot{r}^2 \right\rangle = \frac{a^2\omega^2}{2}\cos^2\theta\,.$$

Equation (7) and condition (8) become

$$\frac{\alpha}{2}\sin(2\varphi_0 - 2\theta) = \sin\varphi_0\,,\ \ \alpha\cos(2\varphi_0 - 2\theta) > \cos\varphi_0\,,\ \ \alpha = \frac{a^2\omega^2}{2gl} \qquad (9)$$

We find the conditions under which it is possible to stabilize the horizontal position of the pendulumm $\varphi_0 = \frac{\pi}{2}$, to be

$$\sin 2\theta = \frac{4gl}{a^2\omega^2}, \quad \cos 2\theta < 0 . \tag{10}$$

## §4 Pendulum oscillations for horizontal oscillations of the suspension point

Let the suspension point make harmonic oscillations with a small amplitude $a$ and big frequency $\omega$ (figure 3).

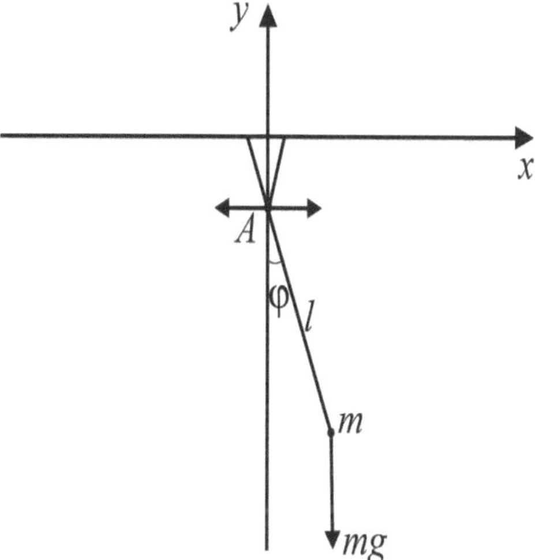


Figure 3

We shall find

$$x = l\sin\varphi + s(t),\quad s(t) = a\sin\omega t,\quad y = -l\cos\varphi .$$

Then we shall find the kinetic and potential energy

$$\mathrm{T} = \frac{m}{2}\left(l^2\dot{\varphi}^2 + 2l\dot{s}\dot{\varphi}\cos\varphi + \dot{s}^2\right),\qquad \Pi = -mgl\cos\varphi$$

and the kinetic potential

$$L = \mathrm{T} - \Pi = \frac{m}{2}\left(l^2\dot{\varphi}^2 + 2l\dot{s}\dot{\varphi}\cos\varphi + \dot{s}^2\right) + mgl\cos\varphi = \\ = \frac{m}{2}\left(\left(l\dot{\varphi} + \dot{s}\cos\varphi\right)^2 + \dot{s}^2\sin^2\varphi\right) + mgl\cos\varphi. \tag{11}$$

We shall find function $\Pi_0(\varphi)$

$$\Pi_0(\varphi) = \left\langle \min_{\dot{\varphi}} L \right\rangle = \left\langle \frac{m}{2}\dot{s}^2\sin^2\varphi + mgl\cos\varphi \right\rangle = \frac{m}{2}\left\langle \dot{s}^2 \right\rangle \sin^2\varphi + mgl\cos\varphi = \\ = \frac{ma^2\omega^2}{4}\sin^2\varphi + mgl\cos\varphi.$$

The extremum of function $\Pi_0(\varphi)$ is found from the first-order condition

$$\frac{d\Pi_0(\varphi)}{d\varphi} = mgl\left(\alpha\sin\varphi\cos\varphi - \sin\varphi\right) = 0, \qquad \alpha \equiv \frac{a^2\omega^2}{2gl} > 0 \tag{12}$$

Function $\Pi_0(\varphi)$ has the critical point $\varphi = 0$. This point corresponds to the lower equilibrium position of the pendulum. Function $\Pi_0(\varphi)$ has at $\varphi = 0$ its maximum under the condition

$$\frac{d^2\Pi_0(\varphi)}{d\varphi^2} = mgl\left(\alpha\cos 2\varphi - \cos\varphi\right) < 0.$$

At $\varphi = 0$ we find the condition of the function maximum $\Pi_0(\varphi)$ which looks like inequality $\alpha < 1$. At $\alpha < 1$ the lower pendulum equilibrium position $\varphi = 0$ will be stable. At $\alpha > 1$ function $\Pi_0(\varphi)$ has its minimum at point $\varphi = 0$ and the lower equilibrium position will be unstable.

At $\alpha = \dfrac{a^2\omega^2}{2gl} > 1$ there are two more equilibrium positions, which come out of the equation

$$\left(\alpha\cos\varphi - 1\right)\sin\varphi = 0, \qquad \alpha = \frac{a^2\omega^2}{2gl} > 0.$$

We shall find these equilibrium positions

$$\cos\varphi = \alpha^{-1}, \qquad \varphi = \pm\arccos\left(\alpha^{-1}\right).$$

The condition for stability of these equilibrium positions on the basis of the minimax criteria

$$\frac{d^2\Pi_0(\varphi)}{d\varphi^2} = mgl\left(\alpha\cos 2\varphi - \cos\varphi\right) = mgl\frac{1-\alpha^2}{\alpha} < 0$$

give the inequality $\alpha > 1$.

We shall model the pendulum movement using differential equations. Studying these we can see the advantages and disadvantages of the *stability minimax criteria.*

Lagrangian equation [1]

$$\frac{d}{dt}\frac{\partial L}{\partial\dot{\varphi}} - \frac{\partial L}{\partial\varphi} = 0$$

will look as follows after transformation

$$\ddot{\varphi} + \frac{1}{l}\ddot{s}\cos\varphi + \frac{g}{l}\sin\varphi = 0.$$

After time averaging of the equation we come to the following equation

$$\ddot{\varphi} + \frac{g}{l}\sin\varphi = 0, \qquad (13)$$

where the influence of the pendulum suspension point oscillations is not seen. Time averaging the Lagrangian function gives,

$$\langle L\rangle = \frac{m}{2}\left(l^2\dot{\varphi}^2 + \langle\dot{s}^2\rangle\right) + mgl\cos\varphi$$

and the Lagrangian equation,

$$\frac{d}{dt}\frac{\partial\langle L\rangle}{\partial\dot{\varphi}}-\frac{\partial\langle L\rangle}{\partial\varphi}=0$$

will look like equation (13).

Before applying the averaging operation we need to re-arrange the pendulum equations of motion to a system of first-order equations.

Let us get down to the canonical differential equation system. We shall introduce the Hamiltonian function

$$H=\frac{\partial L}{\partial\dot{\varphi}}\dot{\varphi}-L$$

from which $\dot{\varphi}$ is eliminated by the following equation

$$p=\frac{\partial L}{\partial\dot{\varphi}}=ml\left(l\dot{\varphi}+\dot{s}\cos\varphi\right).$$

We shall find the value of generalized velocity

$$\dot{\varphi}=\frac{p}{ml^2}-\frac{\dot{s}}{l}\cos\varphi$$

and put in the expression for

$$H=p\left(\frac{p}{ml^2}-\frac{\dot{s}}{l}\cos\varphi\right)-\frac{m}{2}\left(\frac{p^2}{m^2l^2}+\dot{s}^2\sin^2\varphi\right)-mgl\cos\varphi=$$

$$=\frac{p^2}{2ml^2}-\frac{p\dot{s}}{l}\cos\varphi-\frac{m\dot{s}^2}{2}\sin^2\varphi-mgl\cos\varphi\cdot$$

The pendulum equations of motion

$$\dot{p} = -\frac{\partial H}{\partial \varphi}, \qquad \dot{\varphi} = \frac{\partial H}{\partial p}$$

will look as follows

$$\dot{p} = -\frac{p\dot{s}}{l}\sin\varphi + \frac{m\dot{s}^2}{2}\sin 2\varphi - mgl\sin\varphi, \qquad \dot{\varphi} = \frac{p}{ml^2} - \frac{\dot{s}}{l}\cos\varphi .$$

Such equations at $s = a\sin\omega t$ are difficult to solve so we shall execute time averaging of the right parts. As

$$\langle \dot{s} \rangle = \langle a\omega\cos\omega t \rangle = 0, \qquad \langle \dot{s}^2 \rangle = \langle a^2\omega^2\cos^2\omega t \rangle = \frac{a^2\omega^2}{2},$$

then we come to the simplified differential equation system

$$\dot{p} = \frac{ma^2\omega^2}{4}\sin 2\varphi - mgl\sin\varphi, \qquad \dot{\varphi} = \frac{p}{ml^2}. \tag{14}$$

This equation system is canonical with the Hamiltonian function

$$\langle H \rangle = \frac{p^2}{2ml^2} - \frac{ma^2\omega^2}{4}\sin^2\varphi - mgl\cos\varphi .$$

Equation (13) has an integral $\langle H \rangle = const$, or

$$ml^2\frac{\dot{\varphi}^2}{2} - \frac{ma^2\omega^2}{4}\sin^2\varphi - mgl\cos\varphi = const . \tag{15}$$

Equation (15) shows that function $\Pi_0(\varphi)$ is only different from the potential energy in the energy integral by a sign

$$\mathrm{T} + \Pi = const, \quad \mathrm{T} = m\frac{l^2\dot{\varphi}^2}{2}, \quad \Pi = -\frac{ma^2\omega^2}{4}\sin^2\varphi - mgl\cos\varphi .$$

Thus the maximum of the function $\Pi_0(\varphi)$ corresponds to the minimum of the potential energy of the time averaged system of equations for the pendulum oscillations.

Thus the extremum points of function $\Pi_0(\varphi)$ correspond to the stationary points of the system of equations (14).

Stationary points $p=0$, $\varphi=\varphi_0$ are stable if the potential energy $-\Pi_0(\varphi)$ of the time averaged system has its minimum, or if the function $\Pi_0(\varphi)$ has its maximum at the point $\varphi=\varphi_0$.

The minimax criteria for stability does not allow us to research the pendulum motions at all values $\varphi$, but it allows us to find the stationary points of the system of equations (14) without using equations of motion, as well as researching the stability of these stationary points.

We shall show the integral curves of system (14) on the phase plane $(\varphi,\dot{\varphi})$ in figure 4.

I. $0 \le \alpha < 1$

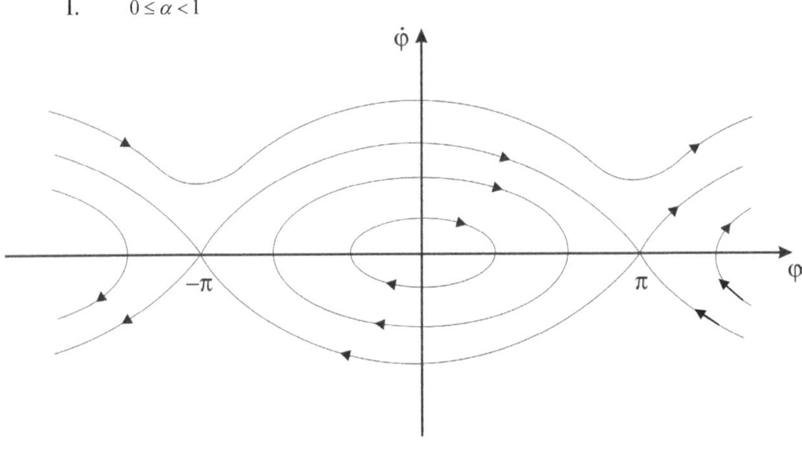

II. $1<\alpha$

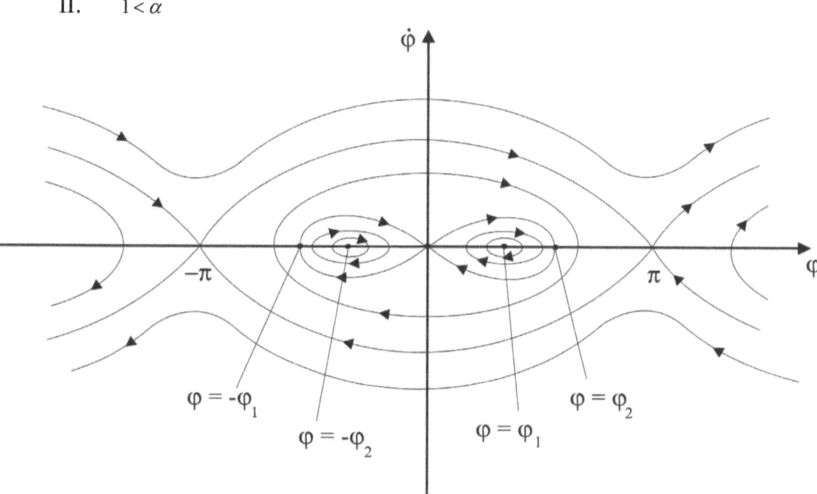


Figure 4

Here $\varphi_1 = \arccos\alpha^{-1}$, $\varphi_2 = \arccos\left(2\alpha^{-1}-1\right)$.

If the pendulum suspension point is oscillating horizontally then at a small frequency of oscillations $\omega$ the pendulum is making oscillations near the lower equilibrium position $\varphi=0$. At an oscillation frequency of more than $\omega = a^{-1}\sqrt{2gl}$ the lower equilibrium position becomes unstable and the pendulum comes to one of the two new equilibrium positions $\varphi = \pm\arccos\left(2gla^{-2}\omega^{-2}\right)$. Further increasing of the the oscillation frequency $\omega$ leads to the equilibrium positions approaching the horizontal position of the pendulum.

## §5 Stability of physical pendulum oscillations at the upper equilibrium position

We shall consider the problem about the stability of the plane oscillations of a body that is flexibly fixed at point $A$, which is making vertical oscillations in the manner

$$y_A = r(t), \quad \langle r(t) \rangle = 0,$$

where $r(t)$ is a periodic function with a small amplitude and big frequency. In this particular case we can suppose

$$r(t) = a \sin \omega t.$$

The body mass is defined as $m$, the body moment of inertia with respect to $B$, where the body's centre of gravity is located, is defined as $I$. The distance from the suspension point of the pendulum $A$ to the centre of gravity $B$ is defined as $l$, the angle of deflection of the line $AB$ from the vertical is defined as $\varphi$. (figure 5).

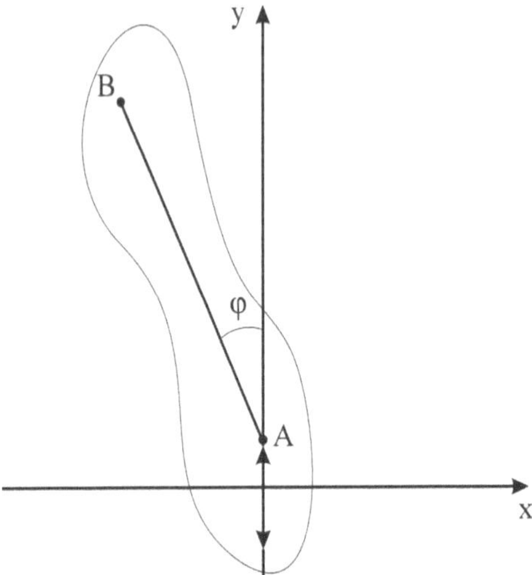


Figure 5

The coordinates of the centre of gravity and the kinetic and potential energy are calculated with the following formulas,

$$x_B = -l\sin\varphi, \qquad y_B = l\cos\varphi + r(t)$$

$$\mathrm{T} = \frac{1}{2}I\dot{\varphi}^2 + \frac{1}{2}m\left(l^2\dot{\varphi}^2 - 2l\dot{r}\dot{\varphi}\sin\varphi + \dot{r}^2\right)$$

$$\Pi = mg\left(l\cos\varphi + r\right).$$

We shall apply the minimax stability criteria for finding the upper equilibrium position's stability $\varphi = 0$. We shall find the following expression

$$\Pi_0(\varphi) = \left\langle \min_{\dot{\varphi}} L \right\rangle = -\frac{m^2l^2\left\langle \dot{r}^2\right\rangle}{2\left(I + ml^2\right)}\sin^2\varphi + \frac{m\left\langle \dot{r}^2\right\rangle}{2} - mgl\cos\varphi \tag{16}$$

Function $\Pi_0(\varphi)$ has its maximum at the point $\varphi = 0$ if the following condition is executed

$$\frac{ml^2}{ml^2+I}\langle\dot{r}^2\rangle > gl\,. \tag{17}$$

At $I=0$ the pendulum's upper stability conditions will coincide with condition (17). The inequality can be written as follows (17)

$$\frac{ml^2}{I_A}\langle\dot{r}^2\rangle > gl\,,$$

where $I_A$ is the inertia moment of the body relatively supporting point $A$.

We shall consider a plain inhomogeneous pendulum with length $L$ and with linear density $\rho(y)$. At this we have the following equalities

$$mL=\int_0^L y\rho(y)\,dy\,,\quad I_A=\int_0^L y^2\rho(y)\,dy\,.$$

The stability condition for the pendulum's upper equilibrium position is expressed with the following inequality

$$\frac{\langle\dot{r}^2\rangle}{g} > \frac{\int_0^L y^2\rho(y)\,dy}{\int_0^L y\rho(y)\,dy}\,. \tag{18}$$

If at $\rho(y)>0$ the improper integral

$$\int_0^\infty y^2\rho(y)\,dy$$

coincides, then the upper equilibrium position of the pendulum of infinite length can be made stable due to the oscillations of the pendulum's supporting point. Anyway, if the linear density of the pendulum $\rho(y)$ is decreasing quickly enough with the pendulum's length, then we can stabilize the upper unstable equilibrium position of a pendulum of a regardless of length $L$.

As an example we shall consider a number of pendulums with the linear change of the plane $\rho(y)$.

For the homogeneous pendulum $\rho = const$ (figure 6.a) the equilibrium condition of the pendulum upper stability position will look like the inequality $\langle \dot{r}^2 \rangle > \frac{2}{3} L$.

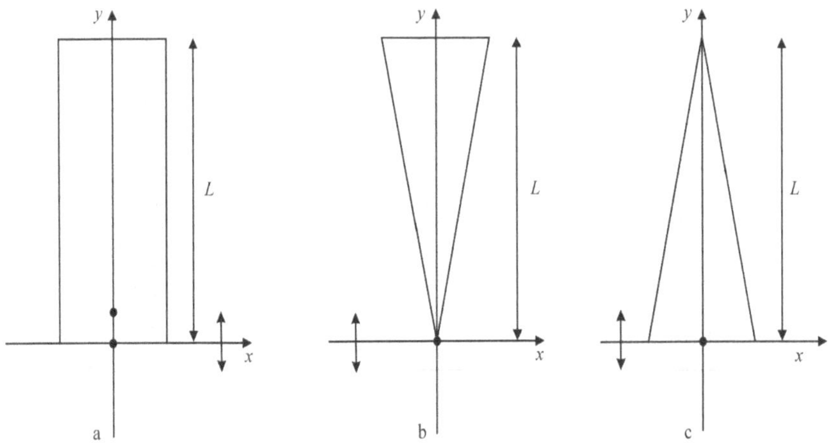


Figure 6

If the density changes in the manner of $\rho = \rho_0 y$ (figure 6.b), then the equilibrium condition for the pendulum's upper stability position comes to the inequality $\langle \dot{r}^2 \rangle > \frac{3}{4} L$.

When the density of the pendulum changes in the manner $\rho = \rho_0 (L - y)$ (figure 6.c), then the equilibrium condition comes to the inequality $\langle \dot{r}^2 \rangle > \frac{1}{2} L$.

The nearer the centre of gravity $B$ is to the supporting point $A$ of the pendulum, the easier it is to stabilise the upper equilibrium position of the pendulum.

## §6 Pendulum vibrating moment

An oscillating effect on the vibrating system can be approximately replaced by the effect of potential forces, with potential

$$\Pi_v = -\left\langle \min_{\dot{q}_i} T \right\rangle . \tag{19}$$

where kinetic energy $T$ is made for the perturbed system.

The generalized vibrating forces are found from the formula [1]

$$F_{vj} = -\frac{\partial \Pi_v}{\partial q_j} = \frac{\partial}{\partial q_j}\left\langle \min_{\dot{q}_i} T \right\rangle \qquad (j = 1,...,n).$$

If function $\Pi_v(q_j)$ has its minimum at some point $q_j = q_{j0} \quad (j = 1,...,n)$, then vibrating forces will tend to move the system into the position $q_j = q_{j0} \quad (j = 1,...,n)$.

We shall consider the pendulum motion only under the influence of vibrating forces when no gravity force exists. We shall suppose that the supporting pendulum point $A$ makes quick oscillations with small amplitude simultaneously in the horizontal and vertical directions in the manner

$$x_A = s(t), \quad y_A = r(t).$$

Mass $m$ of the pendulum with length $l$ is concentrated at point $B$ (figure 7). We shall define the angle between the y-axis and the pendulum as $\varphi$. For the coordinates $x, y$ of point $B$ we shall find the expression

$$x = s(t) - l\sin\varphi, \quad y = r(t) + l\cos\varphi .$$

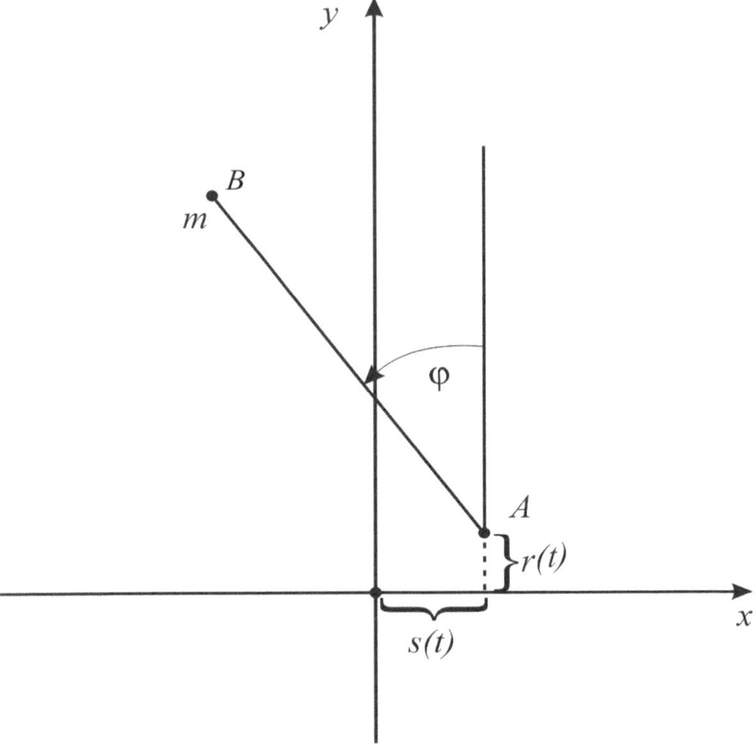


Figure 7

We shall make an expression for the kinetic energy and the potential of the vibrating forces

$$\mathrm{T}=\frac{m}{2}\left(\dot{x}^2+\dot{y}^2\right)=\frac{m}{2}\left(l^2\dot{\varphi}^2-2\dot{\varphi}\,\dot{s}\,l\cos\varphi-2\dot{\varphi}\,\dot{r}\,l\sin\varphi+\dot{r}^2+\dot{s}^2\right)$$

$$\Pi_v=-\left\langle\min_{\dot{\varphi}}\mathrm{T}\right\rangle=\frac{m}{2}\left\langle\left(\dot{s}\cos\varphi+\dot{r}\sin\varphi\right)^2-\dot{r}^2-\dot{s}^2\right\rangle=$$

$$=-\frac{m}{2}\left(\left\langle\dot{r}^2\right\rangle\cos^2\varphi-2\left\langle\dot{s}\dot{r}\right\rangle\cos\varphi\sin\varphi+\left\langle\dot{s}^2\right\rangle\sin^2\varphi\right).$$

Thus, under the influence of vibrations of the supporting point $A$, the vibrating moment $M_v$ appears

$$M_v = -\frac{\partial \Pi_v}{\partial \varphi} = m\left(\left(\langle \dot{s}^2 \rangle - \langle \dot{r}^2 \rangle\right)\frac{\sin 2\varphi}{2} - \langle \dot{s}\dot{r} \rangle \cos 2\varphi\right). \tag{20}$$

The vibrating moment does not appear if

$$\langle \dot{s}^2 \rangle = \langle \dot{r}^2 \rangle, \quad \langle \dot{s}\dot{r} \rangle = 0.$$

We shall consider the case when the pendulum supporting point makes simple harmonic oscillations with amplitude $a$ along the line that makes angle $\theta$ with the y-axis. In this case

$$s(t) = -a\sin\omega t \sin\theta, \quad r(t) = a\sin\omega t\cos\theta.$$

From that we shall find

$$\langle \dot{s}^2 \rangle = \frac{a^2\omega^2}{2}\sin^2\theta, \; \langle \dot{s}\dot{r} \rangle = -\frac{a^2\omega^2}{2}\sin\theta\cos\theta, \; \langle \dot{r}^2 \rangle = \frac{a^2\omega^2}{2}\cos^2\theta$$

and for the vibrating moment we shall find the following expression

$$M_v = \frac{ma^2\omega^2}{4}\sin 2(\theta - \varphi). \tag{21}$$

The vibrating moment $M_v$ (figure 8) tends to overlap with the line along which the pendulum supporting point is oscillating. This line is called the vibration axis.

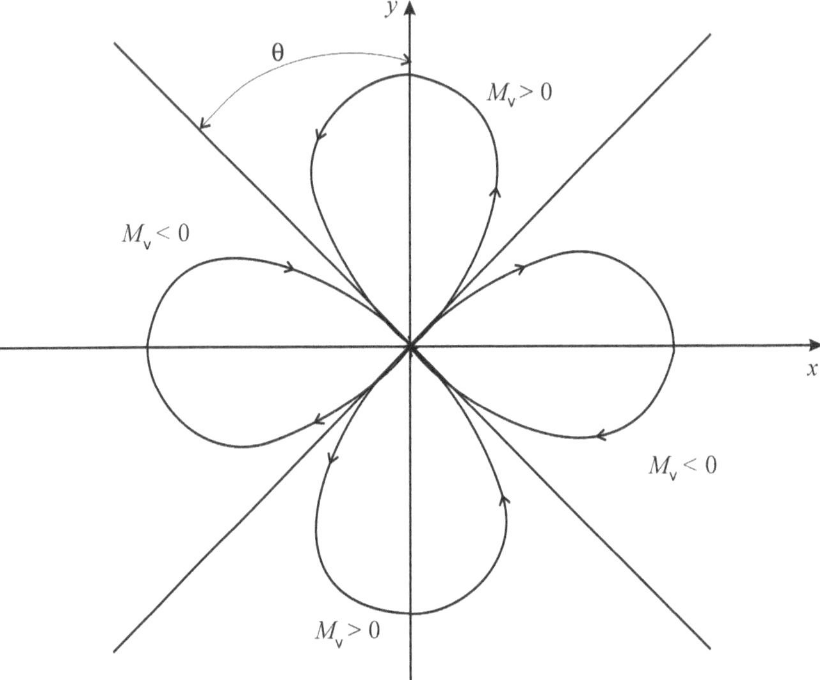


Figure 8

Allowing for gravity forces, the equilibrium position is shifted, but in general it does not coincide either with the upper or with the lower equilibrium position at $\theta \neq k\pi \quad (k = 0, \pm 1)$.

The gravity force moment looks as follows

$$M_g = mgl \sin \varphi .$$

We shall find the conditions under which the pendulum must satisfy to make stable oscillations near the horizontal equilibrium position under the influence of the gravity forces and the vibration of its suspension point. For this, satisfying the following condition is sufficient

$$M_v + M_g = 0, \quad \frac{\partial\left(M_v + M_g\right)}{\partial\varphi} < 0, \quad \varphi = \frac{\pi}{2}.$$

From this we shall come to the received earlier conditions (10)

$$\frac{a^2\omega^2}{4}\sin 2\theta = gl, \quad \cos 2\theta < 0, \tag{22}$$

It can be concluded that $\frac{\pi}{4} < \theta < \frac{\pi}{2}$, $a^2\omega^2 > 4gl$.

The motion of the pendulum will be stable near the horizontal equilibrium position if the condition (22) is satisfied.

We shall notice that in order to satisfy this condition we must also satisfy $a^2\omega^2 > 4gl$, while in order to keep the vertical upper equilibrium position in balance it is sufficient to satisfy the condition $a^2\omega^2 > 2gl$. This phenomenon was confirmed experimentally.

Small oscillations of the pendulum suspension point and the pendulum itself are invisible. The pendulum steadily keeps the horizontal equilibrium position stable. When the frequency or amplitude of the oscillation changes, angle $\varphi$ changes according to the equation (9)

$$\frac{a^2\omega^2}{4}\sin 2\left(\theta - \varphi\right) + gl\sin\varphi = 0. \tag{23}$$

The stability condition for the random equilibrium position $\varphi = const$ looks as follows

$$\frac{a^2\omega^2}{2}\cos 2\left(\theta - \varphi\right) > gl\cos\varphi. \tag{24}$$

Thus, the stabilizing effect of the pendulum upper equilibrium position is only a particular case of the pendulum random stability position, when the vibrating moment and the gravity force moment tend to zero.

At $\varphi = 0$, $\theta = 0$ the equation (23) is satisfied, and the inequality (24) becomes $a^2\omega^2 > 2gl$.

We shall find an expression for the vibrating moment of a physical pendulum, which has mass $m$ and inertial moment $I$ relative to the pendulum's centre of gravity, which is located at point $B$. The distance from the centre of gravity to the suspension point $A$ is equal to $l$. The angle between the vertical and the line $AB$ is defined as $\varphi$. The suspension point $A$ makes simple harmonic oscillations with amplitude $a$ and frequency $\omega$ along the line – the vibration axis, which makes angle $\theta$ with the y-axis (figure 9).

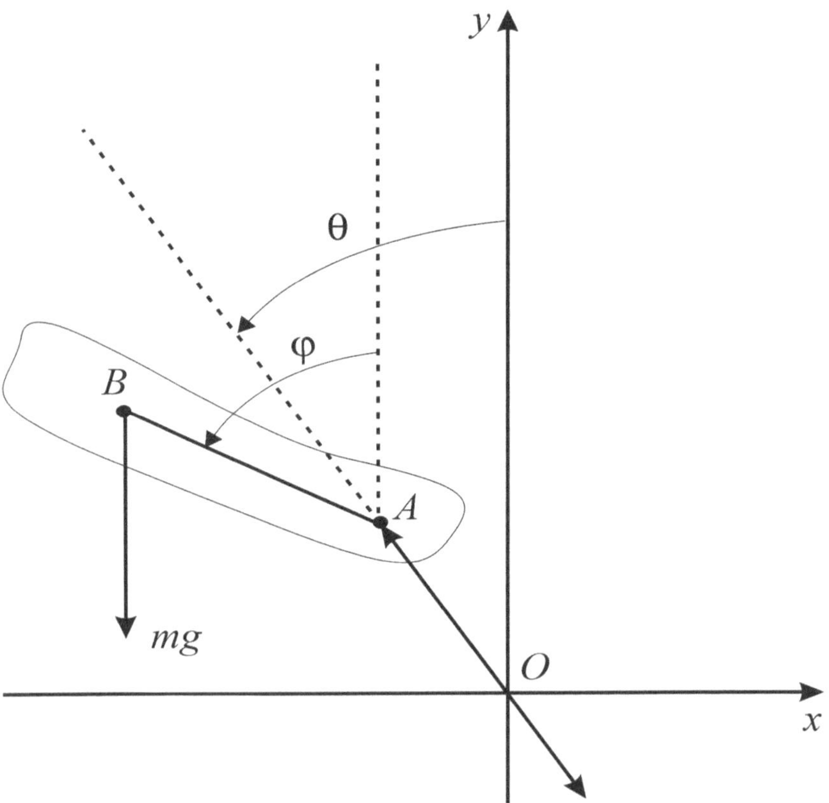


Figure 9

We shall define the co-ordinates of the pendulum's centre of gravity as $x, y$

$$x = -a\sin\omega t\sin\theta - l\sin\varphi, \qquad y = a\sin\omega t\cos\theta + l\cos\varphi.$$

We shall find the following expression for the kinetic energy

$$\mathrm{T} = \frac{1}{2}I\dot{\varphi}^2 + \frac{m}{2}\left(\dot{x}^2 + \dot{y}^2\right) = \frac{I + ml^2}{2}\dot{\varphi}^2 + m\,l\,a\,\omega\,\cos\omega t\cdot\dot{\varphi}\sin\left(\theta - \varphi\right) + \frac{m}{2}a^2\omega^2\cos^2\omega t.$$

Then we shall find the potential of the vibration forces

$$\Pi_v = -\left\langle \min_{\dot{\varphi}} \mathrm{T} \right\rangle = -\frac{1}{2}\left\langle ma^2\omega^2\cos^2\omega t - \left(I_A\right)^{-1} l^2 m^2 a^2\omega^2\cos^2\omega t\sin^2\left(\theta - \varphi\right)\right\rangle =$$

$$= \frac{l^2 m^2}{I_A}\cdot\frac{a^2\omega^2}{4}\sin^2\left(\theta - \varphi\right) - \frac{ma^2\omega^2}{4}; \qquad I_A = I + ml^2$$

and the value of the moment of vibration,

$$M_v = -\frac{\partial \Pi_v}{\partial \varphi} = \frac{l^2 m^2}{I_A}\cdot\frac{a^2\omega^2}{4}\sin 2\left(\theta - \varphi\right). \tag{25}$$

For the moment of the gravitational force we have

$$M_g = mgl\sin\varphi.$$

The steady stability conditions for the physical pendulum at angle $\varphi$ to the vertical will look as follows

$$\begin{aligned} &\frac{l^2 m}{I + ml^2}\cdot\frac{a^2\omega^2}{4}\sin 2\left(\theta - \varphi\right) + gl\sin\varphi = 0, \\ &\frac{l^2 m}{I + ml^2}\cdot\frac{a^2\omega^2}{2}\cos 2\left(\theta - \varphi\right) > gl\cos\varphi. \end{aligned} \tag{26}$$

As follows from these conditions (26), it is easier to stabilize the pendulum whose mass is concentrated at the centre of gravity at $I = 0$, than the one with a distributed mass and the same centre of gravity.

We should notice that stabilization is facilitated by the pendulum's centre of gravity oscillations and it is prevented by the pendulum's oscillations around the centre of gravity.

We shall find the obvious expression for the angle of error $\varphi - \theta$. From formula (26) we get,

$$\varphi - \theta = \frac{1}{2}\arcsin\left(\frac{4gl}{a^2\omega^2}\cdot\frac{I + l^2 m}{l^2 m}\sin\varphi\right). \tag{27}$$

At fixed value $a\omega$ the angle of error $\varphi - \theta$ becomes largest at $\varphi = \frac{\pi}{2}$, i.e. while keeping the pendulum's horizontal position in balance.

## §7 Dynamic stability of the upper equilibrium position of a spherical pendulum

Let the supporting point $A$, of the spherical pendulum make periodic oscillations in vertical and horizontal directions in the manner

$$x_A = \xi(t),\ \ y_A = \eta(t),\ \ z_A = \zeta(t).$$

Here $\xi(t), \eta(t), \zeta(t)$ are periodic functions that are defined by the summations,

$$\xi(t) = \sum_k \xi_k e^{i\omega_k t},\ \ \eta(t) = \sum_k \eta_k e^{i\omega_k t},\ \ \zeta(t) = \sum_k \zeta_k e^{i\omega_k t},$$

where frequencies $\omega_k$ meet the conditions

$$\omega_k \neq 0,\ \ \omega_{-k} = \omega_k,\ \ \min_k |\omega_k| >> 1,$$

and amplitudes $\xi_k, \eta_k, \zeta_k$ are small enough in a module value. The position of point $B$, where mass $m$ is concentrated, is established by coordinates $u, \upsilon$

$$x_B = \xi + u,\ \ y_B = \eta + \upsilon,\ \ z_B = \zeta + \sqrt{l^2 - u^2 - \upsilon^2}.$$

For the kinetic and potential energy we find the expression

$$\mathrm{T} = \frac{m}{2}\left( \left(\dot{u} + \dot{\xi}\right)^2 + \left(\dot{\upsilon} + \dot{\eta}\right)^2 + \left( \dot{\zeta} - \frac{u\dot{u} + \upsilon\dot{\upsilon}}{\sqrt{l^2 - u^2 - \upsilon^2}} \right)^2 \right),$$

$$\Pi = mg\left(\xi + \sqrt{l^2 - u^2 - \upsilon^2}\right). \tag{28}$$

After making transformations we find

$$\left\langle \min_{\dot{u},\dot{\upsilon}} L \right\rangle = \frac{m}{2l^2}\left(\left\langle \dot{\xi}^2 + \dot{\eta}^2 - \dot{\zeta}^2 \right\rangle \left(u^2 + \upsilon^2\right) - \left\langle \dot{\eta}^2 \right\rangle u^2 + 2\left\langle \dot{\xi}\dot{\eta} \right\rangle u\upsilon - \left\langle \dot{\xi} \right\rangle \upsilon^2\right) + \frac{mg}{2l}\left(u^2 + \upsilon^2\right) + \ldots$$

Dots mean the zero-, first-order terms relatively $u,\upsilon$, as well as the third-order terms and the higher-order terms, which were not shown in the formula.

The stability condition is reduced to the existence of a minimum in square form

$$-\Pi_0\left(u,\upsilon\right) = u^2\left(\left\langle \dot{\zeta}^2 \right\rangle - \left\langle \dot{\xi}^2 \right\rangle - gl\right) - 2u\upsilon\left\langle \dot{\xi}\dot{\eta} \right\rangle + \upsilon^2\left(\left\langle \dot{\zeta}^2 \right\rangle - \left\langle \dot{\eta}^2 \right\rangle - gl\right).$$

The necessary and sufficient conditions of the positive function determination $-\Pi_0\left(u,\upsilon\right)$ are reduced to the well-known Silvester inequality

$$\left\langle \dot{\zeta}^2 \right\rangle > gl + \left\langle \dot{\xi}^2 \right\rangle, \quad \left\langle \dot{\xi}^2 \right\rangle > gl + \left\langle \dot{\eta}^2 \right\rangle,$$

$$\left(\left\langle \dot{\zeta}^2 \right\rangle - \left\langle \dot{\xi}^2 \right\rangle - gl\right)\left(\left\langle \dot{\zeta}^2 \right\rangle - \left\langle \dot{\eta}^2 \right\rangle - gl\right) - \left\langle \dot{\xi}\dot{\eta} \right\rangle^2 > 0.$$

Finally the upper equilibrium stability condition for a spherical pendulum will look like the following inequality

$$\left\langle \dot{\zeta}^2 \right\rangle > gl + \frac{1}{2}\left\langle \dot{\xi}^2 \right\rangle + \frac{1}{2}\left\langle \dot{\eta}^2 \right\rangle + \sqrt{0,25\left(\left\langle \dot{\xi}^2 \right\rangle - \left\langle \dot{\eta}^2 \right\rangle\right)^2 + \left\langle \dot{\xi}\dot{\eta} \right\rangle^2}. \tag{29}$$

The inequality (29) shows that vertical oscillations of the supporting point stabilize the upper equilibrium position and the oscillations in horizontal directions remove the stability of the upper equilibrium position.

Let the pendulum supporting point make only plane motion in the x-direction, i.e. $\eta = 0$.

The condition for the upper equilibrium position of a spherical pendulum will look like the following inequality

$$\left\langle \dot{\zeta}^2 \right\rangle > gl + \left\langle \dot{\xi}^2 \right\rangle$$

and it coincides with the stable equilibrium condition for a plane pendulum.

In particular, if the supporting point is making only vertical oscillations then the condition for the pendulum's upper equilibrium position will look like the following inequality

$$\left\langle \dot{\zeta}^2 \right\rangle > gl \, .$$

In the case when the supporting point is making harmonic oscillations

$$\zeta = a \sin \omega t \, ,$$

the stability conditions are reduced to the inequality

$$\frac{a^2 \omega^2}{2} > gl \, ,$$

and are well-known for the plane pendulum.

## §8 Dynamic stability of the equilibrium position for the double pendulum with vertical oscillations of its supporting point

We shall consider the motion of a double pendulum that consists of two weightless rigid rods with lengths $l_1, l_2$. The rods are joined flexibly at point $B$, which has mass $m_1$. At free tag $C$ of the second pendulum there is body $m_2$ fixed (figure 10).

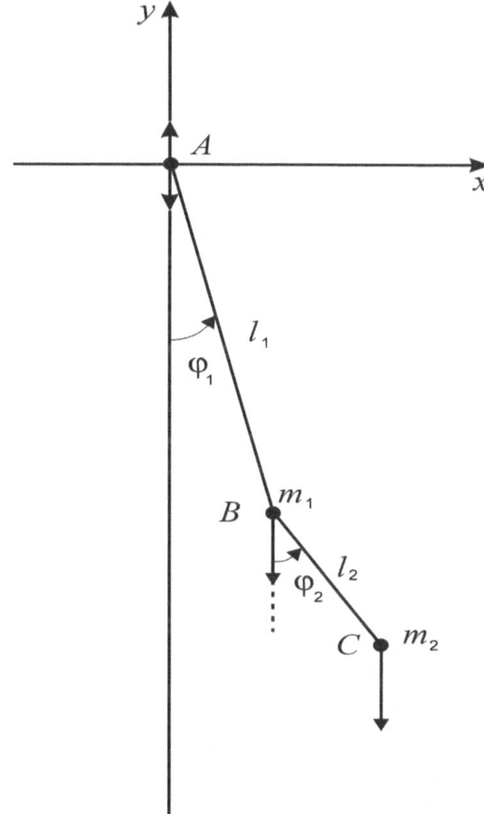


Figure 10

Suspension point $A$ is oscillating vertically in the manner

$$y_A = r(t),$$

where $r(t)$ is a periodic or quasi-periodic function.

We shall define the deflection angles from the vertical as $\varphi_1, \varphi_2$. For the coordinates of points $B, C$ we shall find the expression

$$x_B = l_1 \sin\varphi_1, \quad y_B = r(t) - l\cos\varphi_1;$$

$$x_C = l_1 \sin\varphi_1 + l_2 \sin\varphi_2, \quad y_C = r(t) - l_1 \cos\varphi_1 - l_2 \cos\varphi_2 .$$

For the kinetic and potential energy we shall find the expressions

$$\mathrm{T} = \frac{m_1}{2}\left(l_1^2\dot{\varphi}_1^2 + 2l_1\dot{\varphi}_1\dot{r}\sin\varphi_1 + \dot{r}^2\right) + \frac{m_2}{2}\left(l_1^2\dot{\varphi}_1^2 + l_2^2\dot{\varphi}_2^2 + 2l_1l_2\dot{\varphi}_1\dot{\varphi}_2\cos(\varphi_1 - \varphi_2) + \right.$$
$$\left. + 2\dot{r}\left(l_1\dot{\varphi}_1\sin\varphi_1 + l_2\dot{\varphi}_2\sin\varphi_2\right) + \dot{r}^2\right);$$

$$\Pi = m_1 g\left(r - l_1\cos\varphi_1\right) + m_2 g\left(r - l_1\cos\varphi_1 - l_2\cos\varphi_2\right). \quad (30)$$

We shall apply the minimax stability criteria for researching equilibrium positions. Finding the minimum of the Lagrangian function $L = \mathrm{T} - \Pi$ by variables $\dot{\varphi}_1, \dot{\varphi}_2$, we shall find

$$\left\langle \min_{\dot{\varphi}_1, \dot{\varphi}_2} L \right\rangle = \frac{m_1 + m_2}{2m_1}\left\langle \dot{r}^2 \right\rangle\left(-(m_1 + m_2)\sin^2\varphi_1 + 2m_2\cos(\varphi_1 - \varphi_2)\sin\varphi_1\sin\varphi_2 - m_2\sin^2\varphi_2\right) +$$
$$+ (m_1 + m_2) g l_1 \cos\varphi_1 + m_2 g l_2 \cos\varphi_2 + \ldots .$$

Dots denote terms that do not contain $\varphi_1, \varphi_2$.

The sufficient condition for the function to have a maximum,

$$\Pi_0(\varphi_1, \varphi_2) = \left\langle \min_{\dot{\varphi}_1, \dot{\varphi}_2} L \right\rangle$$

will coincide with the condition of positive determination in square form

$$w(\varphi_1, \varphi_2) = \frac{m_1 + m_2}{2m_1} \langle \dot{r}^2 \rangle \left( (m_1 + m_2)\varphi_1^2 - 2m_2\varphi_1\varphi_2 + m_2\varphi_2^2 \right) +$$
$$+ \frac{m_1 + m_2}{2} g l_1 \varphi_1^2 + \frac{m_2}{2} g l_2 \varphi_2^2 .$$

These conditions will look like the expressions

$$2s + gl_1 > 0; \quad 2s + gl_2 > 0;$$
$$(m_1 + m_2)(2s + gl_1)(2s + gl_2) - 4s^2 m_2 > 0 . \tag{31}$$

where it is denoted

$$s = \frac{m_1 + m_2}{2m_2} \langle \dot{r}^2 \rangle .$$

Different ways of a double pendulum equilibrium come from the lower equilibrium position by changing signs of the values $l_1$, $l_2$. In particular, in the case when the upper equilibrium position of a double pendulum (figure 11) is considered, values $l_1$, $l_2$ in function $w(\varphi_1, \varphi_2)$ are changed into values $-l_1$, $-l_2$. There the condition of positive determination of square form $w(\varphi_1, \varphi_2)$ is reduced to one inequality

$$\langle \dot{r}^2 \rangle > \frac{g}{2} \left( l_1 + l_2 + \sqrt{l_1^2 + l_2^2 + 2l_1 l_2 \frac{m_2 - m_1}{m_2 + m_1}} \right) . \tag{32}$$

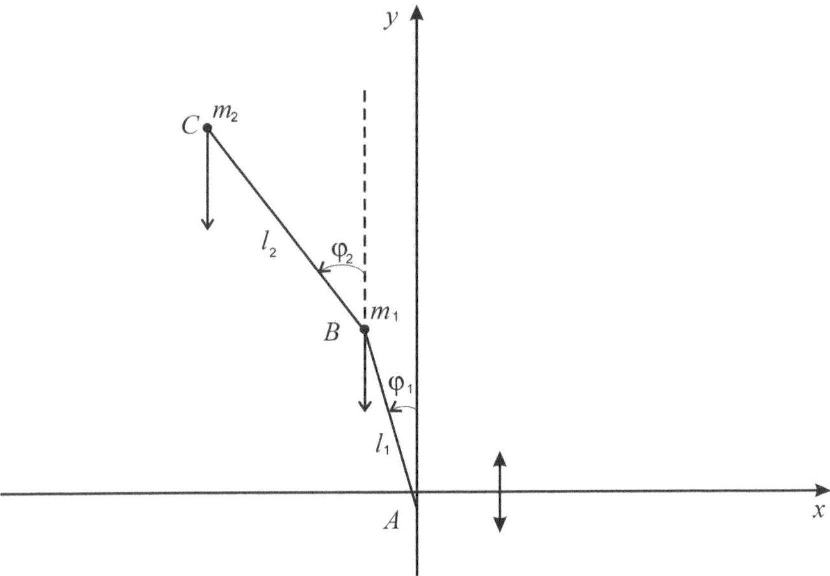


Figure 11

If this inequality is satisfied the upper equilibrium position of the double pendulum will be stable.

For comparison we shall notice that in the case where the double pendulum changes into an ordinary pendulum with one degree of freedom at $\varphi_1 \equiv \varphi_2$, then the condition for the upper equilibrium position will assume the form of the following inequality

$$\left\langle \dot{r}^2 \right\rangle > g \frac{m_1 l_1^2 + m_2 (l_1 + l_2)^2}{m_1 l_1 + m_2 (l_1 + l_2)}. \tag{33}$$

It can be shown that if condition (32) is satisfied, then condition (33) is also always satisfied, i.e. the following inequality is always true

$$\frac{m_1 l_1^2 + m_2 (l_1 + l_2)^2}{m_1 l_1 + m_2 (l_1 + l_2)} \le \frac{1}{2} \left( l_1 + l_2 + \sqrt{l_1^2 + l_2^2 + 2 l_1 l_2 \frac{m_2 - m_1}{m_2 + m_1}} \right)$$

at $m_1 > 0$, $m_2 > 0$, $l_1 > 0$, $l_2 > 0$.

If one of the pendulums change into the zero-length pendulum, for instance $l_2 = 0$, then the condition for dynamic stability of the upper equilibrium position, (32) will assume the form of the following inequality

$$\langle \dot{r}^2 \rangle > g l_1 .$$

Other cases of stabilizing the double pendulum's equilibrium positions are researched in a similar way. Conditions of the dynamic stability of equilibrium positions come from condition (32) by changing the signs of values $l_1, l_2$.

So the stability condition for the equilibrium position, in the case when point $B$ is located above point $A$ and point $C$ is situated below point $B$ (figure 12), assumes the form of the following inequality

$$\langle \dot{r}^2 \rangle > \frac{g}{2}\left( l_2 - l_1 + \sqrt{l_1^2 + l_2^2 - 2 l_1 l_2 \frac{m_2 - m_1}{m_2 + m_1}} \right). \tag{34}$$

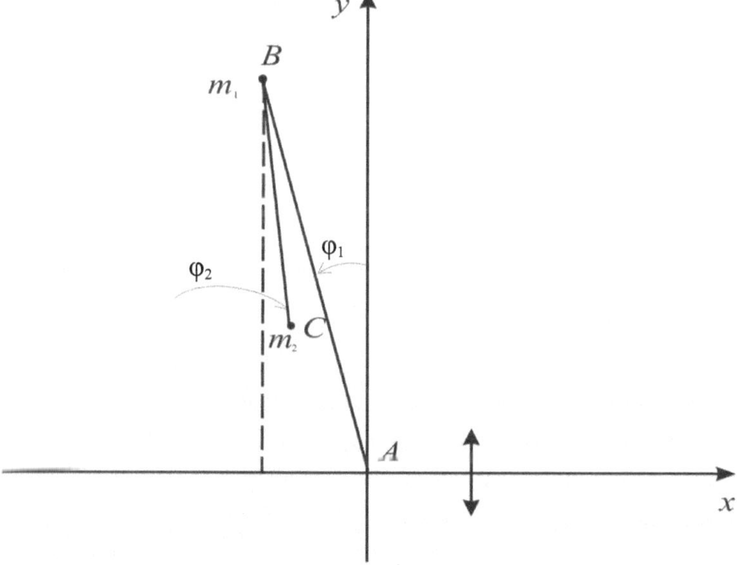


Figure 12

The condition for dynamic stability of another equilibrium position shown in figure 13 assumes the form of the following inequality

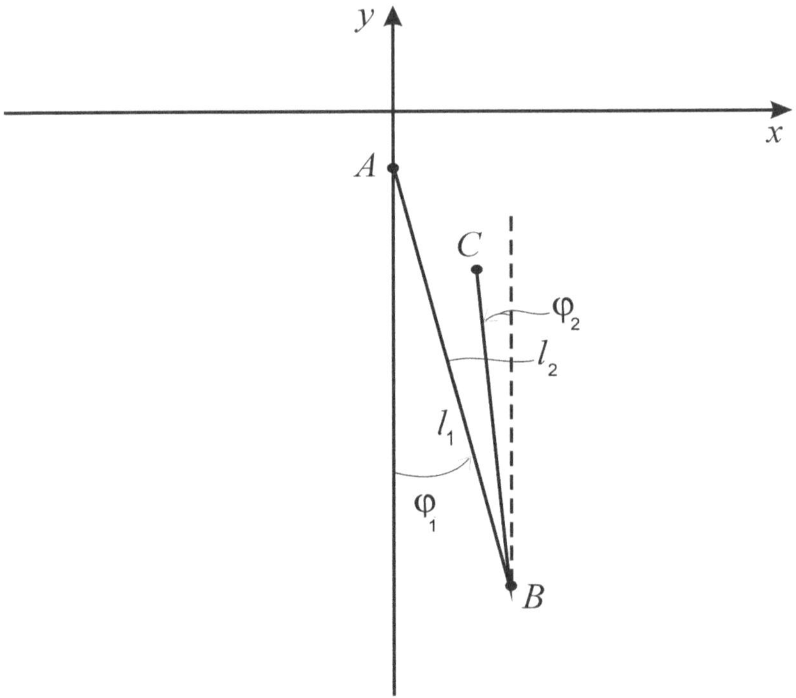


Figure 13

$$\langle \dot{r}^2 \rangle > \frac{g}{2}\left( l_1 - l_2 + \sqrt{l_1^2 + l_2^2 - 2 l_1 l_2 \frac{m_2 - m_1}{m_2 + m_1}} \right) \qquad (35)$$

Of the two equilibrium positions, which are shown in figures 12 and 13, the equilibrium position at which point $C$ takes a lower position is easier to stabilize.

It is interesting to notice that at $l_1 = l_2$ the dynamic stability conditions of the upper and lower equilibrium positions, which are 'folded' in point $B$ of a double pendulum, are equal.

It is easy to make sure that the lower equilibrium position of the double pendulum will always be stable at high-frequency vertical oscillations of the suspension point. In this case the stability condition for the lower equilibrium position assumes the form of the following inequality

$$\langle \dot{r}^2 \rangle > \frac{g}{2}\left(-l_1 - l_2 + \sqrt{l_1^2 + l_2^2 + 2l_1 l_2 \frac{m_2 - m_1}{m_2 + m_1}}\right), \qquad (36)$$

which is always true if the following condition is satisfied $l_1 > 0$, $l_2 > 0$, $m_1 > 0$, $m_2 > 0$.

We shall state that examples of equilibrium of the triple, $n$-multiple pendulum were considered in the work [5]. The stability of the system of joined pendulums was considered in the works [4-7].

## §9 Experimental checking of the existence of the vibration moment existence

In order to check the above stated conclusions about the existence of the vibration moment and it's magnitude, an experimental installation was designed and built. It allows creating harmonic vibration of the pendulum suspension point along the given straight line. [4-6].

The installation contains an eccentric mechanism which is set in rotary motion by an electric engine, straight-line guides which direct the axes of vibration, a worm-gear mechanism of the vibration axis rotation, a photo-electric meter of vibration frequency and an adjustable power source.

The pendulum is a rectangular steel rod with length 100 mm which is flexibly fixed at the suspension point on miniature ball bearings. The accessory axis of bearing allows to use pendulums of different structures. The pendulum's suspension point makes harmonic reciprocating motions along the guides. The law of the pendulum suspension point movement is time sinusoidal with oscillation amplitude $a = 4\,\text{мм}$. The vibration frequency can change smoothly within 1-100 Hz. At a frequency of 100 Hz the acceleration in the construction of flexible parts reaches the value $160\,g \approx 1600\,m/ñ^2$. The electric engine is a collector-type, a direct-current motor, with a rate of power of 20 W. The position of the vibration axis can be changed smoothly within $0-360^{\circ}$. The installation scheme is shown in figure 14.

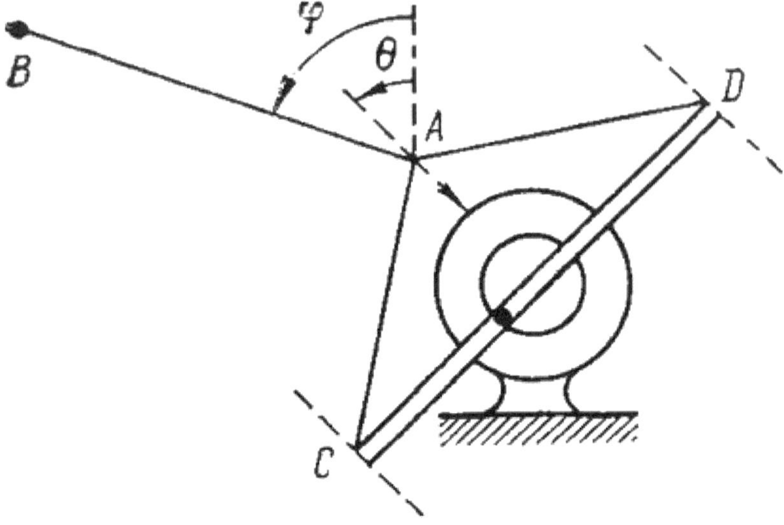


Figure 14

The eccentric installed on the engine's axis gives reciprocating motion to plate $ACD$, which can move freely along the straight-line guides which are shown in figure 14 with a dotted line. At that, point $A$ in which the pendulum is flexibly fixed is making ordinary harmonic oscillations along the straight-line guides.

Figure 15 shows this installation during the experiment when the pendulum takes a stable horizontal position. The pointer on the scale shows the direction of the vibration axis along which the pendulum's suspension point is oscillating.

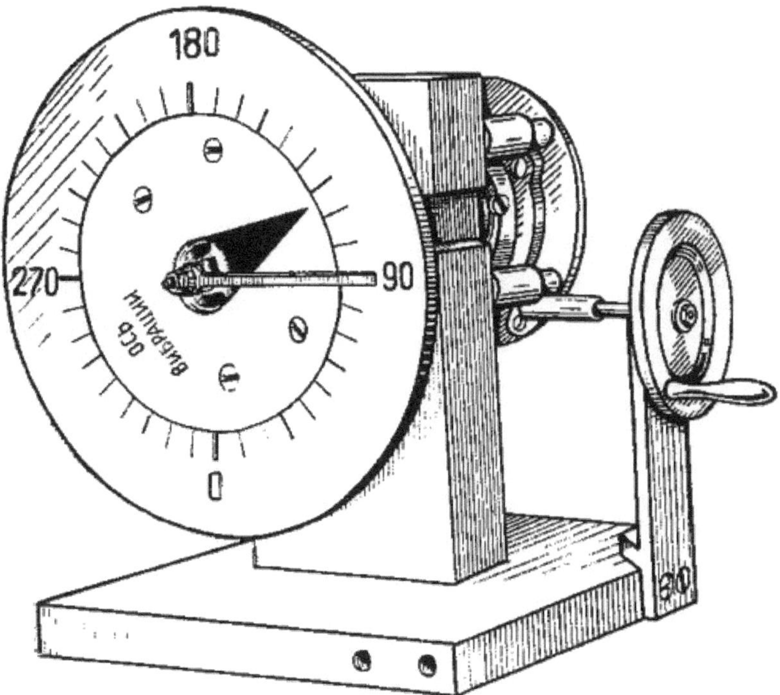


Figure 15

At large enough values of the vibration frequency and slow changes of the direction of the axis the pendulum follows the direction of the axis of vibration with some error angle. The error angle decreases while the frequency of vibration increases. When the pendulum is deflected from the position of stability it comes back to the initial stable position whilst oscillating.

The same experimental installation showed the stable oscillations of a double pendulum, in particular, in the case when the second pendulum made a rotary motion.

## §10 Averaging method in canonical systems

We shall consider a system of canonical differential equations

$$\frac{dp_j}{dt} = -\mu\frac{\partial H}{\partial q_j}; \quad \frac{dq_j}{dt} = \mu\frac{\partial H}{\partial q_j} \qquad (j=1,\dots,n), \tag{37}$$

where $H = H(t, p_s, q_s, \mu)$ is the Hamilton function, which is analytical relatively $p_s, q_s, \mu$, $\mu$ is a small parameter , time $t$ is periodic or quasi-periodic. The averaging methods were created in the works of L. Euler, K. Gauss, L. Lagrange, P. Laplace and improved in the works by K. Delone, M. Lindtsted, K. Gilden, A. Poincare. These methods were mainly applied for solving problems of celestial mechanics which use canonical equations. A new stage in the development of averaging methods began in the works by B. Van der Paul, L.I. Mandelshtam and N.D. Papalexi, who researched oscillating solutions of uncanonical differential equations. A big contribution in the popularizing of averaging methods was made by mathematicians from Kiev, such as N.N. Bogolyubov and Y.A. Mitropolsky. The most important up-to-date results on averaging methods were stated in the works of T.G. Stryzhak [4-6]. An an averaging operation with the aid of analytical extension of images by Laplace functions was generalized.

From time averaging the equation system (37) it assumes the form of a first order approximation

$$\frac{dp_j}{dt} = -\mu\frac{\partial\langle H\rangle}{\partial q_j}, \quad \frac{dq_j}{dt} = \mu\frac{\partial\langle H\rangle}{\partial p_j} \qquad (j=1,\dots,n), \tag{38}$$

where it is denoted

$$\langle H\rangle = \lim_{\mathrm{T}\to\infty}\frac{1}{\mathrm{T}}\int_0^{\mathrm{T}} H(t, p_s, q_s, \mu)\,dt.$$

In the works by Delone, Lindstedt and Poincare it was shown that changing from the equation system (37) to the autonomous equation system (38) is a first approximation of variable substitution which is built in rows. These rows were used in solving problems of celestial mechanics. A. Poincare showed that the applied rows are, as a rule, divergent but have an asymptotic character at $\mu \to 0$.
Let the equation system assume the shape of the following

$$\frac{dp_j}{dt} = -\frac{\partial H}{\partial q_j}, \quad \frac{dq_j}{dt} = \frac{\partial H}{\partial p_j}, \quad H = H(\omega t, p_s, q_s)\,(j, s = 1, \dots, n). \tag{39}$$

After the substitution $\omega t = \tau$ we come to the canonical equation system

$$\frac{dp_j}{d\tau} = -\frac{1}{\omega}\frac{\partial H_0}{\partial q_j}, \quad \frac{dq_j}{dt} = \frac{1}{\omega}\frac{\partial H_0}{\partial p_j},$$

$$H_0 = H(\tau, p_s, q_s) \;\; (j, s = 1, \dots, n) \tag{40}$$

where we can substitue $\mu = \omega^{-1}$ and the equation system (40) can be time averaged and assumes the following shape

$$\frac{dp_j}{dt} = -\frac{\partial \langle H \rangle}{\partial q_j}, \quad \frac{dq_j}{dt} = \frac{\partial \langle H \rangle}{\partial p_j} \tag{41}$$

where it is denoted

$$\langle H \rangle = \lim_{T \to \infty} \frac{1}{T} \int_0^T H(\omega t, p_s, q_s)\, dt.$$

Thus, the presence of a large frequency $\omega$ in the equation system (39) is equivalent to the presence of a small parameter $\mu$ in the equation system of (37).

For systems of equations (38) and (41) conclusions about stability can be made from the presence of the energy integral.

$$\langle H \rangle = const.$$

## §11 Proof of the Minimax stability criteria for a single-degree-of-freedom system

Firstly, we shall consider oscillations of a single-degree-of-freedom system, as it is more difficult to understand the main idea of the proof for the system with $n$ degrees of freedom.

Let the Lagrangian function assume the following form

$$L(t,\dot{q},q)=a(t,q)\dot{q}^2+b(t,q)\dot{q}+c(t,q). \tag{42}$$

We shall pass to the canonical equation system by introducing a new variable – the impulse

$$p=\frac{\partial L}{\partial \dot{q}}=a(t,q)\cdot 2\dot{q}+b(t,q).$$

Excluding $\dot{q}$ out of the Hamiltonian function we shall get

$$H=p\dot{q}-L=\frac{p^2}{4a}-\frac{b}{2a}p+\frac{b^2-4ac}{4a}.$$

Of all the considered examples the following equality was satisfied

$$\left\langle\frac{b}{2a}\right\rangle=0, \tag{43}$$

so the following equality will be true

$$\langle H\rangle=\left\langle\frac{1}{4a}\right\rangle p^2+\left\langle\frac{b^2-4ac}{4a}\right\rangle. \tag{44}$$

The stable stationary point $(0;q^*)$ of the differential equation system of motion

$$\dot{p} = -\frac{\partial \langle H \rangle}{\partial q} = -p^2 \frac{\partial}{\partial q}\left\langle \frac{1}{4a} \right\rangle - \frac{\partial}{\partial q}\left\langle \frac{b^2 - 4ac}{4a} \right\rangle,$$

$$\dot{q} = \frac{\partial \langle H \rangle}{\partial p} = 2p\left\langle \frac{1}{4a} \right\rangle$$

is defined from the conditions

$$p = 0, \quad \left\langle \frac{b^2 - 4ac}{4a} \right\rangle = \min_q \left\langle \frac{b^2 - 4ac}{4a} \right\rangle.$$

Function $-\Pi_0 = \left\langle -\frac{b^2 - 4ac}{4a} \right\rangle$ has a minimum at the point $q^*$.

We shall apply the minimax stability criteria. We shall find

$$\min_{\dot{q}} L(t, \dot{q}, q)$$

from the equation

$$p = \frac{\partial L}{\partial \dot{q}} = a \cdot 2\dot{q} + b = 0 .$$

We shall find the value of the minimum of $L$

$$\min_{\dot{q}} L = a\left(-\frac{b}{2a}\right)^2 + b\left(-\frac{b}{2a}\right) + c = \frac{4ac - b^2}{4a}$$

and the expression for the function

$$\Pi_0(q) = \left\langle \frac{4ac - b^2}{4a} \right\rangle,$$

which has a maximum at point $q = q^*$. It proves the validity of the minimax criteria of stability.

## §12 Derivation of the minimax stability criteria in the general case

Let the Lagrangian function which describes the motion of the system with $n$ degrees of freedom assumes following shape.

$$L(t,\dot{q}_j,q_j)=Q^*A(q_j)Q+B(t,q_j)\dot{Q}+c(t,q_j),\quad (j=1,...,n), \tag{45}$$

where $A(q_j)$ is a symmetric, positive definite matrix, elements of which depend on the generalized coordinates $q_1,...,q_n$, $B(t,q)$ is a line vector which meets the following condition

$$\langle B(t,q_j)\rangle \equiv 0 \quad (j=1,...,n). \tag{46}$$

Vector components $B(t,q_j)$ and function $c(t,q_j)$ are periodic or quasi-periodics functions of $t$.

The column vector component $\dot{Q}$ has generalized velocities $\dot{q}_j$ $(j=1,...,n)$. Vector $\dot{Q}^*$ is transposed relatively to vector $\dot{Q}$.

The motion of the system is described by the Lagrangian equations

$$\frac{d}{dt}\frac{\partial L}{\partial \dot{q}_j}-\frac{\partial L}{\partial q_j}=0,\quad (j=1,...,n).$$

We shall pass from the Lagrangian equations to the canonical system of differential equations

$$\dot{p}_j=-\frac{\partial H(t,p_k,q_k)}{\partial q_j},\quad \dot{q}_j=\frac{\partial H(t,p_k,q_k)}{\partial p_j}\quad (j=1,...,n),$$

where the Hamiltonian function $H(t,p_k,q_k)$ is defined by the formula

$$H(t,p_k,q_k)=\sum_{j=1}^{n}p_j\dot{q}_j-L(t,\dot{q}_j,q_j),\qquad p_j=\frac{\partial L}{\partial \dot{q}_j}. \tag{47}$$

We shall introduce a column vector $P$, components of which are impulses $p_1,\dots,p_n$. From equation (45), (47) we shall receive the following vector equality

$$P^*=\frac{\partial L}{\partial \dot{Q}}=2\dot{Q}^*A(q_j)+B(t,q_j),\qquad (j=1,\dots,n).$$

Excluding vector $\dot{Q}$ from the Hamiltonian function (47) we come to the obvious expression for $H$

$$H(t,p_j,q_j)=\frac{1}{4}P^*A^{-1}(q_j)P-\frac{1}{4}P^*A^{-1}(q_j)B^*(t,q_j)-\frac{1}{4}B(t,q_j)A^{-1}(q_j)P-$$
$$-\frac{1}{4}B(t,q_j)A^{-1}(q_j)B^*(t,q_j)-c(t,q_j),\qquad (j=1,\dots,n).$$

We shall find the average value of function $H(t,p_j,q_j)$ by time

$$\langle H(t,p_j,q_j)\rangle=\frac{1}{4}P^*A^{-1}(q_j)P+\left\langle\frac{1}{4}B(t,q_j)A^{-1}(q_j)B^*(t,q_j)-c(t,q_j)\right\rangle,\qquad (j=1,\dots,n).$$

We shall introduce the auxiliary function

$$\Pi_0(q_j)=-\left\langle\frac{1}{4}B(t,q_j)A^{-1}(q_j)B^*(t,q_j)-c(t,q_j)\right\rangle,\qquad (j=1,\dots,n)$$

which depends only on the generalized coordinates $q_1,\dots,q_n$.

If function $\Pi_0(q_j)$ has a maximum at some point $q_j=q_j{}^*$, $(j=1,\dots,n)$, then the Hamiltonian function $\langle H(t,p_j,q_j)\rangle$ has at some point $p_j=0$, $q_j=q_j{}^*$ a minimum and the stationary solution
$p_j=0,\quad q_j=q_j{}^*,\quad (j=1,\dots,n)$

of the Hamiltonian equation system

$$\dot{p}_j = -\frac{\partial\langle H\rangle}{\partial q_j}, \quad \dot{q}_j = \frac{\partial\langle H\rangle}{\partial p_j}, \qquad (j = 1,...,n) \tag{48}$$

is stable and coincides with the stable periodic or almost periodic solution of the system (48).

Finding Hamiltonian function $H(t, p_j, q_j)$ can take long calculations. So we shall show a simpler way of finding the function $\Pi_0(q_j)$.

We shall find the minimum of the Lagrangian function $L(t, \dot{q}_j, q_j)$ by generalized velocities $\dot{q}_j$ $(j = 1,...,n)$. Finding $\dot{q}_j$, $(j = 1,...,n)$ from the equation system

$$\frac{\partial L}{\partial \dot{Q}} = 2\dot{Q}^* A(q_j) + B(t,q)_j = 0, \qquad (j = 1,...,n)$$

and excluding $\dot{Q}$ from the Lagrangian function we shall find

$$\min_{\dot{q}_j} L(t, \dot{q}_j, q_j) = -\frac{1}{4} B(t,q_j) A^{-1}(q_j) B^*(t,q_j) + c(t,q_j)\,\dot{q}_j, \qquad (j = 1,...,n).$$

We shall average this expression by time $t$ and find the following function

$$\Pi_0(q_j) = \left\langle \min_{\dot{q}_j} L(t, \dot{q}_j, q_j) \right\rangle. \tag{49}$$

Finally we come to the following minimax stability criteria.

If a time-average $t$ from the minimum by generalized velocities $\dot{q}_j$ of Lagrangian function $L(t, \dot{q}_j, q_j)$ has maximum by generalized coordinates $q_j$, then this maximum coincides with the stable periodic or almost periodic solution of the system.

# Appendix

**Topics for self-studying:**

1. Generalize the minimax stability criteria for the infinite number of degrees of freedom. Study fluid tilt in a vessel.

2. Develop the minimax stability criteria for the system with random oscillations, in particular for Markovian random processes.

3. Develop the minimax instability criteria for oscillations.

4. Repeat the experiment with stabilization of pendulums (double, triple)

5. Make a children toy that is based on the property of a spherical pendulum and rises at vertical oscillations of its supporting point.

6. As the minimax stability criteria is not always suitable for researching the stability, develop the improved criteria for stability in cases where conditions (43), (46) are not satisfied.

## Literature


1. Лурье А.Н. Аналитическая механика. – Москва: Физматгиз, 1961. – 824 с. (Lurie A.N. Analytical Mechanics. – Moscow: Physmathgiz, 1961.- 824 pp.)

2. Hirsch P. Das Pendel mit aszillierendem Aufhängenpunkt. – Zeitschrift für angewandte Mathematik und Mechanik. 1930. Bd. 10. S. 41-52.

3. Erdelyi A. Uber die kleinen Schwingungen eines Pendels mit oszillierendem Aufhängenpunkt. – Zeitschrift für angewandte Mathematik und Mechanik. 1934 Bd. 14. Heft 4. S. 235-247.

4. Стрижак Т.Г. Метод усреднения в задачах механики. – Киев-Донецк, "Вища школа", 1982. – 250 с. (Stryzhak T.G. Averaging Method in Mechanics Problems. – Kiev-Donetsk, "Vyshcha shkola", 1982. 250 pp.)

5. Стрижак Т.Г. Методы исследования динамических систем типа "маятник". – Алма-Ата, "Наука", 1981. – 256с., 1982. – 250 с.) (Stryzhak T.G. Research Methods of Dynamic Systems of a "Pendulum" Kind. – Alma-Ata, "Nauka", 1981. – 256 pp., 1982. – 250 pp.)

6. Стрижак Т.Г. Асимптотический метод нормализации. – Киев, "Вища школа", 1984. – 280 с. (Stryzhak T.G. Asymptotic Method of Normalization. – Kiev, "Vyshcha shkola", 1984. – 280 pp.)

7. Валеев К.Г. Динамическая стабилизация неустойчивых систем. – Изв. АН СССР. Механика твердого тела. 1971, №4, с.13-21. (Valeev K.G. Dynamic Stabilization of Unstable Systems. – News of Academy of Science USSR. Solid body mechanics. 1971, #4, pp.13-21.)

8. Блехман И. И. Вибрационная механика. – Москва: “Физмат”, 1994. – 400 с. (Blekhman I. I. Vibration Mechanics. – Moscow: “Fizmat”, 1994. – 400 pp.)

Project ***"Modern Mathematics for Engineers"*** includes publishing the following works:

1 Difference equations with random coefficients
2 Stability of solutions of differential equations systems with random coefficients
3 Random values modeling
4 Optimal control synthesis
5 Principle of reduction
6 New method of averaging
7 New determinant theory
8 Minimax criterion of stability
9 Numerical methods of stability research
10 Analytical functions from matrix
11 Frequently criteria of stability

## The Editors

**"Modern Mathematics for Engineers"**
**First IAESTE trainee students**

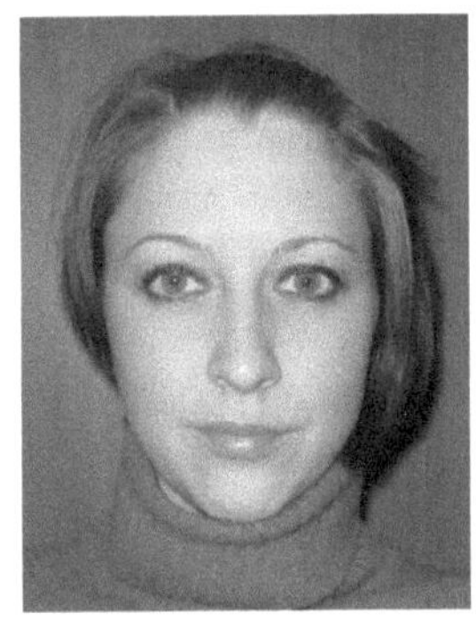

**Mirjana Vukelja**
**Switzerland**
**ETH Zurich**

**Bettina Carina Sieber**
**Germany**
**TU Munich**

**Laura Lambertz**
**Germany**
**RWTH Aachen**

**Ki Yeun Kim**
**USA**
**Carnegie Mellon University**
**Department of Mathematical Sciences**

**Peter Smith**
**UK**
**Queen's University Belfast**
**School of Mathematics and Physics**

**Heide Gieber**
**Austria**
**Vienna University of Technology**
**Department of Economical Mathematics**

**Sarah Burden**
**UK**
**University of St. Andrews**
**Department of Mathematics**
**(Prof. E. Priest)**

**Andrea Trautsamwieser**
**Austria**
**Technical University of Vienna**
**Department of Mathematics**

**Leonard Neuhaus**
**Germany**
**Ludwig - Maximilians University of Munich**
**Department of Physics**

**Erkan Koc**
**Germany**
**University of Bonn**
**Department of Computer Science**

**David Ellis,
Imperial College,
London, UK**

**Henning Hans Petzka, RWTH
Aachen, Germany**

**Leonard Neuhaus, Sarah Burden,
Andrea Trautsamwieser, Erkan Koc**

*ibidem*-Verlag

Melchiorstr. 15

D-70439 Stuttgart

info@ibidem-verlag.de

www.ibidem-verlag.de
www.ibidem.eu
www.edition-noema.de
www.autorenbetreuung.de

Zeitfracht Medien GmbH
Ferdinand-Jühlke-Straße 7
99095 Erfurt, Deutschland
produktsicherheit@kolibri360.de